SCHOOL PUBLISHERS

Orlando   Austin   New York   San Diego   Toronto   London

Visit *The Learning Site!*
www.harcourtschool.com

Introduction

After you paint a picture on a sheet of paper, the paint dries. The colors stay where you put them.

Think about what you see when you watch a program on television. The pictures keep changing. The colors don't stay in the same place.

How does a television set show so many different colors on its screen? Actually, a television screen only seems to have hundreds of different colors. In fact, all those colors are made by dots of red, green, and blue light.

Rainbows

Thinking about sunlight can help us understand colors. Sunlight looks white to the eye. White is a light, bright color.

What you usually can't see is that sunlight is made up of many colors. White has all of the colors of the rainbow mixed together.

You see sunlight's true colors when there is a rainbow. A rainbow appears when sunlight shines in and out of raindrops. The sun is behind you when you face a rainbow.

As a ray of light passes through a raindrop, the ray bends. When the ray of light bends, its colors spread apart and you can see them. It's hard to believe so many colors can hide in white sunlight!

The colors in a rainbow are always in the same order. Violet is nearest to the inside of the bow. Then you see blue, green, yellow, orange, and finally red.

A rainbow in the sky is just one way that sunlight separates into its different colors. You can also see a rainbow when sunshine hits the spray of a waterfall or a lawn sprinkler. A prism or a cut diamond can also break apart white light.

In all of these ways, you can see that white light is really a blend of all of the colors of light.

Circles of Light

A rainbow spreads white light apart into colors. The circles of light shown in the picture below put colors together into white light. In the picture, three different circles of light overlap one another.

The three circles show the three primary colors of light. Primary colors of light are the three colors you can put together and get white. They are red, green, and blue.

This can be confusing, because there are two different kinds of primary colors. Remember, red, green, and blue are the three primary colors for *light.* They aren't the same as the three primary colors for *paint.*

When the three primary colors of light mix, they form different colors. All three primary colors together make white.

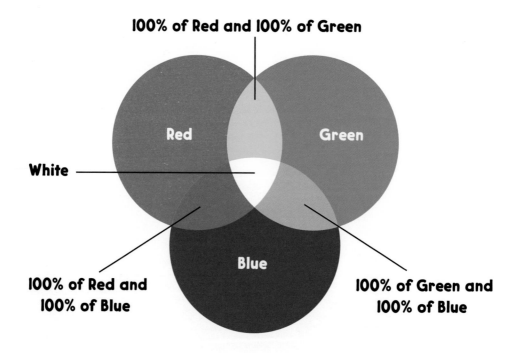

100% of Red and 100% of Green

Red   Green

White

Blue

100% of Red and 100% of Blue

100% of Green and 100% of Blue

Look at the picture of the circles of light. A mix of red and blue light makes magenta. Mixing blue and green makes cyan. Adding green to red makes yellow.

Putting together primary colors to make new colors is called *adding colors.* If you add all three primary colors, you get white. That's why the middle area, where all three circles cross, is white.

The circles of light tell you what happens when you mix the same amount of each primary color. What happens if you use a different amount of each primary color? If you mix a lot of red light with a little green, you won't get pure yellow.

Wouldn't it be fun to shine different amounts of colored lights, just to see what you would get?

You could come up with any color that your eyes can see. To get black, though, you would have to shine no light at all.

By mixing the primary colors of light, we can make all of the other colors that our eyes can see.

How You See Color

When you watch a program on television, you see colors of all kinds. How can your eyes see so many different colors?

Light shines out from the television set, and some of it enters your eyes. In the back of each eye, the light touches two kinds of special cells. These are the cells that sense light.

These special cells are called rods and cones because of their shape.

Rods allow you to see in dim light, such as when you're outdoors at night. Try standing or walking where there isn't much light. You'll find that you see things in black and gray instead of in color. Rods don't sense color. Rods see things that are a little to the side instead of straight ahead.

Cones work only in medium and bright light, such as in a room with the lights on or outdoors in daylight. Cones help you see clearly straight ahead. When you read, you're using your cones.

Cones are special because they can sense colors. However, each cone can sense only one color. This is where the three primary colors come in.

Your eyes have millions of cones. One kind senses red light, one kind senses blue, and the third kind senses green. They can be called red, blue, and green cones.

Suppose that you are watching television, and you see a picture of a blue sky. When the blue light from the

The cones of your eyes see all of these colors as mixtures of red, green, and blue.

television reaches your eyes, your blue cones sense the light and send a "blue" message to your brain.

If there is white snow in the television picture, what happens? Remember that white light has red, blue, and green light in it. Your blue cones sense blue, your red cones sense red, and your green cones sense green.

Your brain gets a message that this light has all three primary colors in it. The light is white.

Sometimes no light at all falls on the cones, so you see black. Your cones send messages to your brain only in red, green, and blue, but your brain adds them together to see many colors.

Connecting the Dots

It is your brain's job to understand what your eyes see. If your eyes see a shape bouncing across the grass, your brain decides whether the shape is a puppy or a rubber ball.

If your cones sense only red from a traffic light, your brain knows that the light is red and that you shouldn't cross the street.

Your brain notices even small differences between colors. You can tell yellow-green from blue-green. You know light blue from dark blue. With only three kinds of cones, you see hundreds of different colors and shades of color.

Your brain has another trick. You can see only part of something and still know what the thing is. If you saw part of a bicycle behind some bushes, you would know

Your brain notices even small differences in what you look at. If you see part of something, you know what it is.

that the object was a bicycle. If you saw a friend in a crowd of people, you would know him or her, even if you saw only part of the friend's face.

You can look at a picture made of dots and see a shape. This is easier when there are a lot of dots. Your brain "connects the dots" and sees a whole picture.

In fact, you probably couldn't stop yourself from seeing the shape. That's how your brain works.

This is amazing. Your eyes have millions of cones, and your brain gets a message from every one. Your brain connects millions of dots to see what is around you.

You don't see spaces between the cones because your mind connects their messages into shapes and colors.

When you look at these dots, your brain connects them and sees a picture.

Color Television

What if a machine could make a picture the same way your brain sees one? What if it could blend red, blue, and green light to make one color from three?

That's what a television set does.

A television doesn't blend three colors in one big place, like the circles of light shown on page 4. It sends color messages from a lot of smaller places, just as the cones of your eyes do.

Think about what would happen if you tore a picture out of an old magazine and cut it into little pieces. If you tried, you could cut the pieces so small that each tiny bit would have only one color on it.

A color television screen is covered with tiny boxes called *pixels.* The word *pixel* is short for PICture ELement. A pixel is the smallest piece that makes up a television picture.

Each pixel has a red dot, a green dot, and a blue dot. A pixel is very small, and its dots are close together.

When a television is on, an energy beam shoots at the pixels to make them light up. The beam shoots differently at each pixel.

In one pixel, the energy beam might aim at the red dot and not at the blue or green. In that case, only the red light would shine. In another pixel, the energy

In every pixel on a television screen, three colors can blend together to make the one color we see. All of these pixels together give us a television picture.

beam might hit all three colored dots, sending out equal amounts of red, blue, and green light. Then the pixel would look white.

You see only one color for each pixel, because your brain adds the colors together. This works very well because the cones of your eyes can sense only red, green, and blue light. The television needs to shine only red, green, and blue light to tell your eyes which colors to see.

All pixels are small and close together, but there are spaces between them. Why do you see the picture and not see the spaces? Your brain "connect the dots" when you see only part of something. Looking at the television screen, you see shapes and colors, instead of just many dots.

Computer Monitors

The words and images you see on your computer monitor are also made up of pixels. A typical computer monitor can display 480,000 pixels. This is more than most televisions can display, which means that a computer monitor has a sharper picture than a television has. You probably don't notice this, however, because images on televisions move more than they do on computer monitors.

Televisions are being developed that have much sharper images than computer monitors have. These television screens have 2,073,600 pixels!